X-TREME FACTS: SCIENCE

TECHNOLOGY

by Catherine C. Finan

BEARPORT PUBLISHING

Minneapolis, Minnesota

Credits:
Cover and title page, © Ociacia/Shutterstock, © FabrikaSimf/Shutterstock, and © ekka-pon/Shutterstock; 4, Everett Historical/Shutterstock.com; 5 top, George Munger/Public Domain; 5 right, Bass Otis/Public Domain; 5 middle, Library of Congress/Public Domain; 5 left, David P. Lewis/Shutterstock.com; 6 top, James Hoban/Public Domain; 6 left, John Christian Rauschner/Public Domain; 6 right, John Trumbull/Public Domain; 7 top, George Cooke/Public Domain; 7, .E. Walker/ Public Domain; 7 left, right, Gilbert Stuart/Public Domain; 8 bottom background Joraca/Shutterstock; 8 bottom left, 8-29 Austen Photography; 8 top, Suzanne Tucker/Shutterstock; 8 right, LightField Studios/Shutterstock; 9 top, The White House/Public Domain; 9 left, Nationaal Archief/Creative Commons; 9 right, (Mr) Smith, War Office official photographer/Public Domain; 9 bottom background, AudeVivere/Creative Commons;10 top, Executive Office of the President of the United States/Public Domain; 10, LightField Studios/Shutterstock; 11 top, Ritu Manoj Jethani/Shutterstock.com; 11 middle, Sashkin/Shutterstock; 11, Executive Office of the President of the United States/Public Domain; 12 top, Carol M. Highsmith/General Services Administration/Public Domain; 12, Leika production/ Shutterstock; 13 top, Annie Leibovitz/Released by White House Photo Office/Public Domain; 13 middle, Official White House Photo by Chuck Kennedy/Public Domain; 13 bottom, padmak/Shutterstock.com; 13 right, Everett Historical/Shutterstock; 14, Bumble Dee/Shutterstock.com; 14 right, Chernetskaya/Dreamstime. com; 15 top, 15 bottom, 27, Pete Souza, Executive Office of the President of the United State/Public Domain; 15 middle, Ronald Regan Archives/Public Domain; 16, Executive Office of the President of the United States/Public Domain; 17 top, 1000 Words/Shutterstock.com; 17 right, Records of the White House Photo Office/Public Domain; 17 left, Viorel Sima/Shutterstock; 17, Orhan Cam/Shutterstock; 18 top, Everett Historical/Shutterstock; 18, The U.S. National Archives/Public Domain; 19 top, Orhan Cam/Shutterstock; 19 top left, bottom left, 21 bottom middle, LightField Studios/Shutterstock; 19 top right, Nicole Glass Photography/ Shutterstock.com; 19 bottom right, Gage Skidmore/Creative Commons; 20 top, Joseph Sohm/Shutterstock.com; 20 right, Harriet Anderson Stubbs Murphy/ Public Domain; 21 top, 360b/Shutterstock.com; 21 bottom left, Roman Tiraspolsky/Shutterstock.com; 21 bottom right, Roman Tiraspolsky/Shutterstock.com; 22 top, U.S. Navy photo by Photographer's Mate 2nd Class Daniel J. McLain/Public Domain; 22 bottom, Executive Office of the President of the United States/ Public Domain; 23 top, DoD photo by U.S. Air Force Staff Sgt. Marianique Santos/Public Domain; 23 bottom, Official White House Photo by Joyce N. Boghosian/ Public Domain; 23 bottom left, Gelpi/Shutterstock; 23 bottom right, Anna Om/Shutterstock; 24 top, The White House. Samantha Appleton, photographer/Public Domain; 24 left, Nik Merkulov/Shutterstock; 24 right, Gift; Michael Geissinger, photographer/Public Domain; 25 top, Official White House Photo by D. Myles Cullen/Public Domain; 25 middle, Brian A Jackson/Shutterstock; 25 bottom left, VGstockstudio/Shutterstock; 25 bottom right, White House Photographic Office/ Public Domain; 26 top left, Anders Zorn/Public Domain; 26 top right, Anders Zorn/Public Domain; 26, Official White House Photo by Joyce N. Boghosian)/Public Domain; 26 bottom right, StockImageFactory.com/Shutterstock; 27 top, U.S. Government/

President: Jen Jenson
Director of Product Development: Spencer Brinker
Editor: Allison Juda

Produced for Bearport Publishing by BlueAppleWorks Inc.
Managing Editor for BlueAppleWorks: Melissa McClellan
Art Director: T.J. Choleva
Photo Research: Jane Reid

Library of Congress Cataloging-in-Publication Data

Names: Finan, Catherine C., 1972- author.
Title: Technology / Catherine C Finan.
Description: Minneapolis, Minnesota : Bearport Publishing Company, [2021] | Series: X-treme facts: science | Includes bibliographical references and index.
Identifiers: LCCN 2020042960 (print) | LCCN 2020042961 (ebook) | ISBN 9781647476779 (library binding) | ISBN 9781647476847 (paperback) | ISBN 9781647476915 (ebook)
Subjects: LCSH: Science—Juvenile literature. | Technology—Juvenile literature.
Classification: LCC Q163 .F43 2021 (print) | LCC Q163 (ebook) | DDC 600—dc23
LC record available at https://lccn.loc.gov/2020042960
LC ebook record available at https://lccn.loc.gov/2020042961

For more information, write to Bearport Publishing, 5357 Penn Avenue South, Minneapolis, MN 55419.
Printed in the United States of America.

Contents

Let's Talk Tech

When you hear the word *technology*, you might think of computers, tablets, and smartphones. But these popular devices are not the whole story! Technology is anything that uses science to solve a problem. It often makes possible what was once thought to be impossible. From ancient structures to modern medical marvels, technology is terrific!

A Japanese **supercomputer** performs **415 quadrillion computations in a single second!**

The technology used to build the pyramids of Giza in Egypt is still impressive—more than 4,500 years later!

Speedy maglev trains use magnets to keep them floating just above the track!

3D printers can make new body parts out of living human cells!

A tiny but high-tech camera can be swallowed in a **capsule** to show doctors what's happening inside your intestines!

People who are afraid of technology have technophobia!

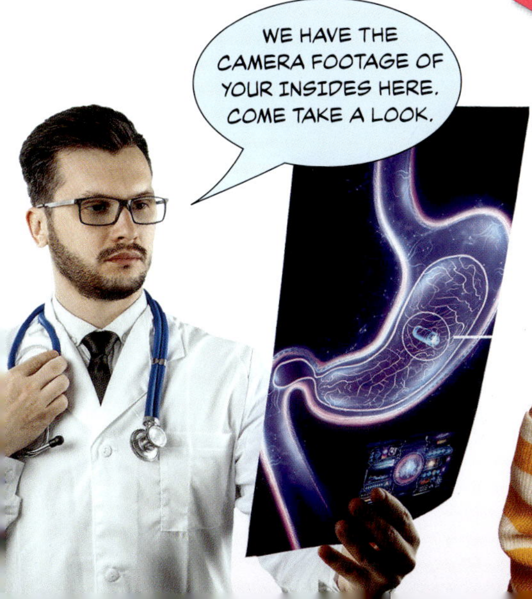

Compute This!

The computer has changed our world, and we have changed the computer—a lot! Today's tiny versions look almost nothing like the first house-sized computers. And they are almost everywhere. Tiny but powerful computers are inside our phones, appliances, and most of the modern **gadgets** we use every day! Think about how many homes you would need to fit all of your devices if tech hadn't gone tiny!

Some of the biggest tech companies, including Apple, Microsoft, Google, and Amazon **all started in garages!**

The first general-purpose computer was called ENIAC. It took up about 1,500 square feet (139 sq m).

Today, a single **microchip** the size of your fingernail can do more than ENIAC ever could!

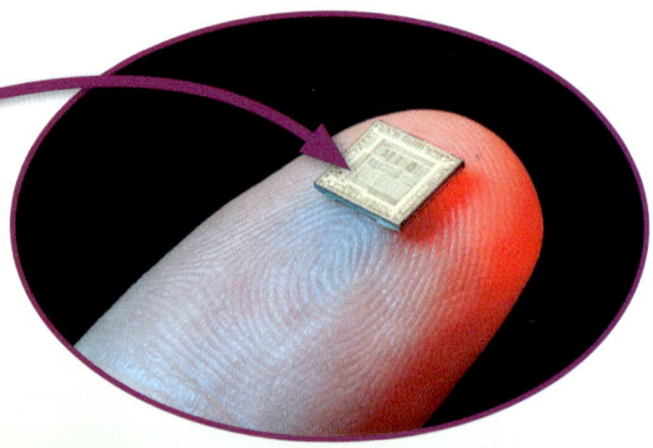

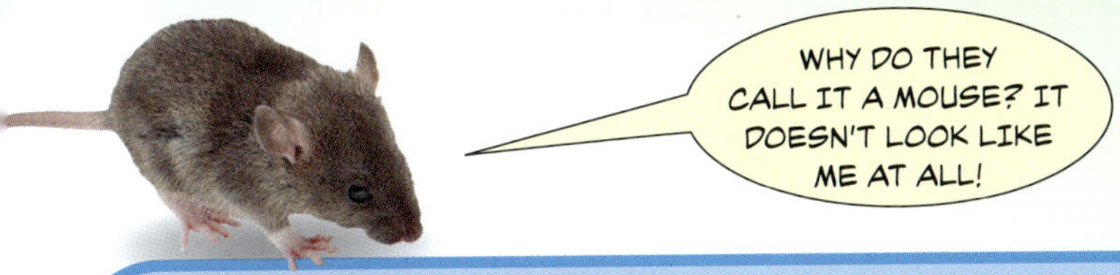

The first computer mouse was made of wood!

The world's **smallest computer could fit on a grain of rice!**

A woman named Ada Lovelace wrote the first computer program in the 1840s!

Home Is Where the Tech Is

Beyond computers, tablets, and smartphones, technology is everywhere we look. Let's start in your home. Your TV, lights, refrigerator, and even the faucet—all of these are technological **innovations**. We take them for granted now, but not so long ago none of these things existed! In the future, who knows what fantastic devices we'll have in our homes. Robots that can cook a meal and clean up after? A flying car parked in the driveway? The sky's the limit!

Solar panels use the sun's energy to heat and power homes.

No snoozing! **There's a smart alarm clock that makes you chase it around to turn it off!**

Ancient Romans didn't have faucets—they got their water from aqueducts.

A new gadget uses **ultrasonic** waves to turn food scraps on your dirty dishes into plant soil.

Monitors can detect your pets and let you soothe them when you are away from home.

The High-Tech Toilet

In your home technology tour, you might overlook one of the greatest inventions ever—the toilet! It took many technological tweaks over time to arrive at today's flush toilet. But now we're looking beyond that. Now some toilets do much more than just flush the yuck away. What's next for the trusty toilet?

In 1596, Sir John Harington may have invented the first flush toilet for Queen Elizabeth I.

In the 1880s, **Thomas Crapper** (yes, Crapper!) made the first widely used flush toilet.

Some toilets today are heated, self-cleaning, and self-**deodorizing**—and they save on water, too!

Bored while in the bathroom? How about a toilet that plays music? It exists!

World Toilet Day is observed each year on November 19.

That's Electrifying

Using the bathroom in the dark can get a little . . . messy . . . so hopefully you keep the lights on. Light bulbs work because of electricity—the flow of tiny charged particles. Controlling electricity has made life easier and more fun. Today, electricity powers everything from your bedside lamp to your TV to your computer. Electricity is excellent!

When the metal key on Benjamin Franklin's kite sparked during a thunderstorm, he proved that lightning was a form of electricity.

Electric current is measured in volts. It was named after Alessandro Volta, who invented the first electric battery in 1799.

Electricity can be made from **fossil fuels**, rushing water, sunlight, or wind. But it can also be made from burning *poop*!

Some electricity-producing wind turbines have blades as long as a football field!

Nikola Tesla invented electricity first, but Thomas Edison took the credit. Shocking!

Now We're Getting Somewhere

Electricity keeps our world moving. Increasingly, many of our vehicles are powered by electricity rather than fuel. But no matter where you travel, you're using some form of technology to get there. From the invention of the wheel to modern bicycles, trains, cars, and airplanes, transportation has always relied on technology to get going.

Some military aircraft can fly faster than the speed of sound!

Some modern cars don't need a driver. A computer does all the driving!

Submersibles can withstand the crushing pressure of the deep ocean!

A U.S. Navy deep-diving submarine traveled to the deepest point on Earth—the Mariana Trench!

A maglev train in Japan has gone up to 374 mph (603 kph)!

Leonardo da Vinci had the idea for a helicopter more than 420 years before the first one was made!

Space: The Final Frontier

We've seen transportation technology at work on land and in the air. But what about in outer space? Humans have done what was once thought impossible—they've traveled beyond Earth's **atmosphere**! The first **satellite** was launched in 1957. The first people walked on the moon in 1969. Today, people live in space aboard the International Space Station (ISS). What other super space voyages does the future hold? Strap in and hold on!

In 2004, the **rovers** Spirit and Opportunity began exploring Mars and sending information and images back to Earth.

In 1610, the astronomer Galileo discovered four of Jupiter's moons with a new telescope he made himself.

Aboard the ISS, special technology turns astronauts' pee into drinking water! *Mmmm* . . . refreshing!

People may be traveling to Mars as early as 2024! The trip could take half a year.

Some say the first permanent human settlement on Mars could be up and running by 2028!

Dress for Tech-cess

If you're going into space, you're going to need something to wear. Actually, whether you're going to an extreme environment or just out for a walk in the park, technology has you covered! From the coldest temperatures to the hottest, from the crushing pressure of the deep sea to the **vacuum** of space, tech has helped us stay safe and comfortable. Just choose your destination or activity, and technology will provide you with the proper attire!

Today's space suits can handle temperatures ranging from 250 degrees Fahrenheit (121 degrees Celsius) to –250°F (–157°C).

Neil Armstrong took his famous first steps on the moon wearing a space suit created by a bra maker!

Smart workout clothes let you adjust your music's volume and tell you when to tweak your yoga pose!

Some sleeping bags keep you warm even at 0°F (−17.8°C). That's some pretty hot tech!

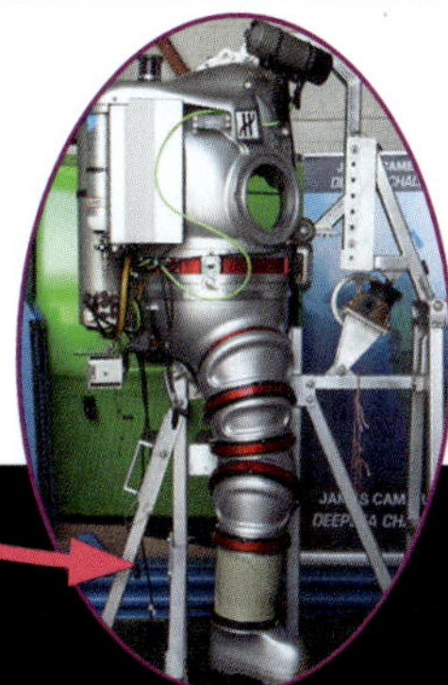

The Exosuit lets you explore up to 1.2 miles (2 km) below the ocean's surface.

Wearable technology that uses cameras and computers **helps U.S. soldiers make smart choices in battle.**

Teeny-Tiny Tech

Sometimes the technology we wear is smaller than a speck of dust! Microscopic pieces of silver kill the fungi and bacteria that can make our clothes smell. **Nanotechnology** works with atoms and molecules—the smallest building blocks of matter. It allows us to make products that can absorb X-rays, scatter light, keep water out, and control heat—all on a super-small scale. But the possibilities for this technology are anything but teeny!

The *nano* in nanotechnology stands for nanometer. **There are one million nanometers in a single millimeter!**

In the future, **nanomachines inside people's bodies might repair damage** and fight sickness!

Just one of your hairs is about 80,000 to 100,000 nanometers wide!

Some new fabrics are coated with nanowhiskers—tiny fibers that dirt can't pass through. This keeps the fabrics' deeper layers clean.

NASA scientists have suggested making an elevator from Earth to space using tiny nanotubes!

Sunscreens coat your skin with nanoscopic materials that block the sun's harmful rays.

Mind-Blowing Medicine

A nanomachine delivering medicine directly into our bodies sounds cool! Especially since it wasn't so long ago that the recommended cure for headaches was to drill holes into the skull. *Yikes!* As doctors learned more about the human body and developed new technology, people began to enjoy longer, healthier lives. Today, with everything from 3D-printed body parts to nanorobots attacking cancer cells, modern medical technology is totally mind-blowing!

Modern **imaging** machines take super-detailed pictures of our insides!

Alexander Fleming discovered the **antibiotic** penicillin by accident! Now, it easily cures infections that once killed people!

An **antibody** in llamas blocks the COVID-19 virus from entering the animals' cells. Who knew our furry friends could help create a **vaccine**!

Researchers are working on a tiny chip that tells your skin cells to heal injuries.

Without modern medicine, you'd live to be about 45—the average length of life during the Middle Ages!

Get Real with Virtual Reality

Many of the doctors who keep us healthy are being trained with almost nothing. Virtual reality (VR) places them inside a 3D experience, allowing them to learn and practice within **simulated** surroundings. VR is used to train not only doctors but also pilots, soldiers, and astronauts. Oh, and it can be used in some pretty fun games. What other worlds are waiting to be explored through VR?

Surgeons can practice operations with the help of VR. Practice makes perfect!

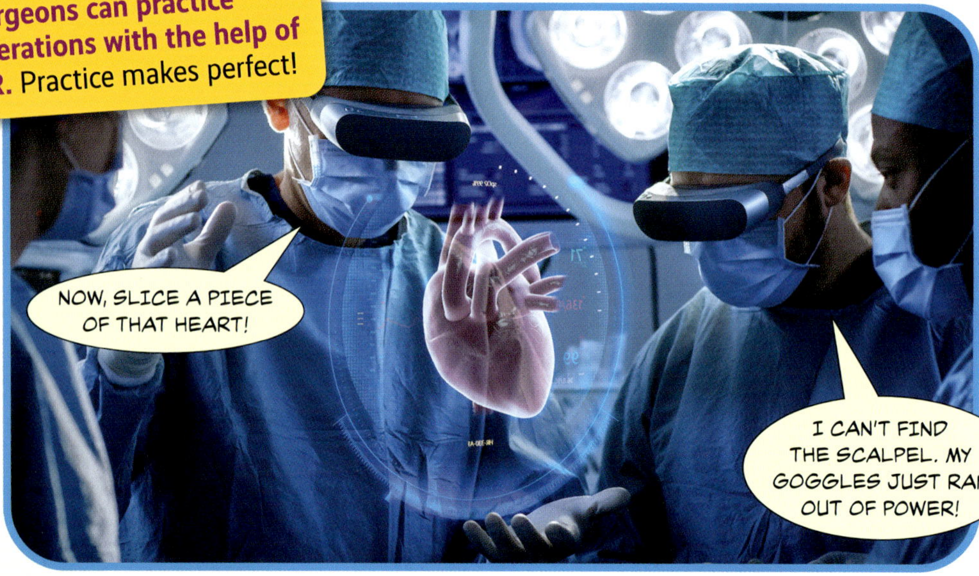

VR is being used to train pilots and even airplane mechanics!

VR lets you dock cargo vehicles on the ISS and go on a virtual space walk, too!

Do you want to walk on the moon? VR can take you there!

Technology of Tomorrow

Technology has taken us to places that once seemed impossible. Things like smartphones, robots, nanotechnology, and VR used to be the stuff of science fiction, but now they're everywhere. And the more we learn, the faster technology continues to develop. What exciting developments are just around the corner? With the wonders of technology, just about anything is possible!

Artificial intelligence (AI) lets computers copy the way humans think and learn.

Someday soon, you might be able to control your computer with just your thoughts!

A drone is a flying robot that can be preprogrammed or guided by a remote control.

Drones deliver packages and food to people's doorsteps. One even delivered a kidney to a hospital for a transplant operation!

Personalized cancer vaccines may soon allow your **immune system** to spot and kill cancer cells on its own.

Sophia is a **humanoid** robot that can talk, move, and show emotions. **She can even sing and draw!**

Coin Battery
Activity

Batteries use a chemical reaction to make electricity. Wet cell batteries were the very first rechargeable batteries. They are made from two types of metal and a liquid acid. The most common example of a wet cell battery is a regular car battery. It's time to make your own wet cell battery!

We use batteries every day. They come in all shapes and sizes, from big ones in cars to tiny ones in cell phones.

What You Will Need

- A pencil
- 9 pennies
- Cardstock
- Scissors
- A bowl of vinegar
- Aluminum foil
- Scotch tape

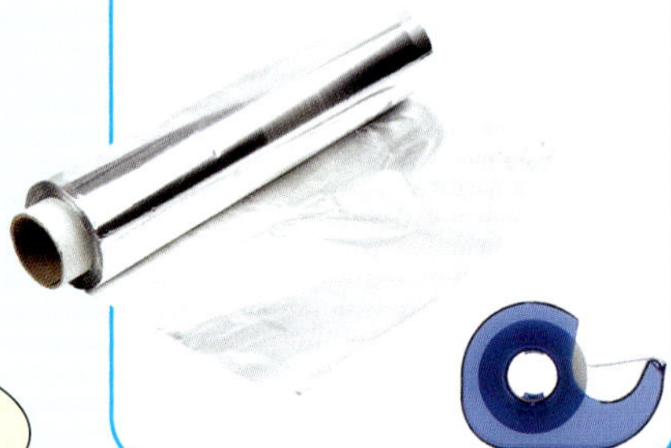

Step One

Use a pencil to trace around a penny nine times on the piece of cardstock. Cut out each circle.

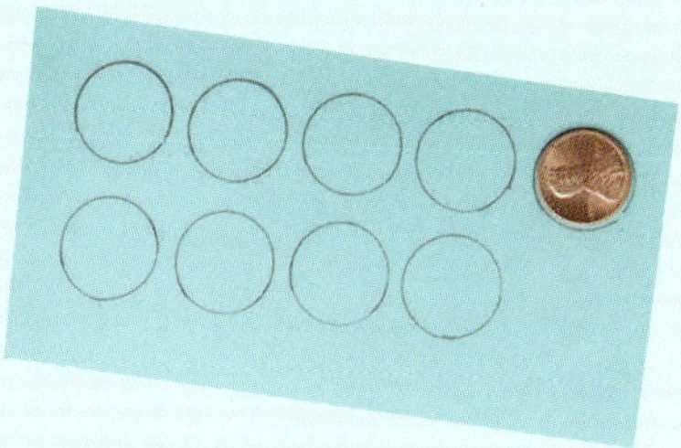

Step Two

Place the circles in the bowl of vinegar.

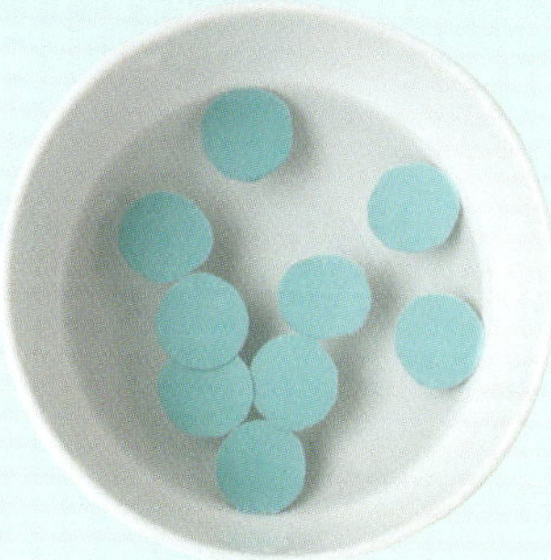

Step Three

Trace around a penny onto a piece of aluminum foil nine times. Cut out each circle.

Step Four

Place a vinegar-soaked cardstock circle on top of a penny. Place an aluminum foil circle on top of the stack.

Step Five

Continue stacking pennies, cardstock, and foil in that order until all circles and coins have been placed. Wrap a piece of tape around the stack.

Step Six

Moisten your fingertips with vinegar. Hold the top and bottom of the stack between your moistened fingers. Feel a little buzz? You have made a battery!

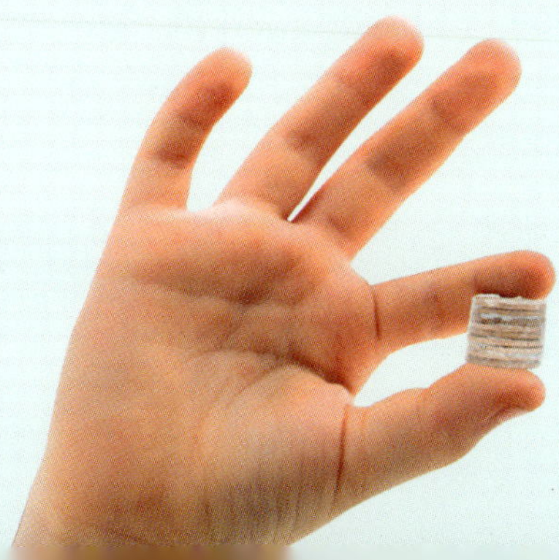

Glossary

antibiotic medicine used to stop the growth of disease-causing bacteria

antibody a protein made by the body that attaches itself to harmful germs to stop infections

aqueducts structures that look like bridges and carry water over valleys

atmosphere the layer of gases that surrounds Earth

capsule a tiny case

deodorizing getting rid of or hiding the odor of something

fossil fuels energy sources, such as coal, oil, and gas, that are made from the remains of plants and animals that died millions of years ago

gadgets small tools that help do a particular job

humanoid looking or acting like a human

imaging creating a picture with the help of tools, such as cameras, computers, X-rays, magnets, or scanners

immune system the system that protects your body from sickness and disease

innovations new ideas, products, or ways to do something

microchip a tiny plate that holds many electronic parts; microchips are used to run computers

monitors devices for collecting information

nanotechnology the science of working with atoms and molecules to make tools and devices that are extremely small

rovers vehicles made to travel on the surface of a planet or moon in order to collect information

satellite a spacecraft that is sent into orbit around a planet

simulated made to seem like something real

submersibles vessels capable of working underwater

supercomputer a large, very fast computer used mostly for calculations

ultrasonic relating to sound beyond the range that can be heard by humans

vaccine a medicine that protects people against a disease

vacuum a space empty of all matter

Read More

Bethea, Nikole Brooks. *The Internet (Super Science Feats).* Minneapolis: Pogo, 2019.

Nagelhout, Ryan. *Freaky True Stories About Technology (Freaky True Science).* New York: Gareth Stevens Publishing, 2017.

Rathburn, Betsy. *Artificial Intelligence (Epic: Cutting-Edge Technology).* Minneapolis: Bellwether Media, 2021

Learn More Online

1. Go to **www.factsurfer.com**
2. Enter "**X-treme Technology**" into the search box.
3. Click on the cover of this book to see a list of websites.

Index

About the Author

Catherine C. Finan is a writer living in northeastern Pennsylvania. She enjoys writing about a wide range of subjects, including technology—but she admits she's a little suspicious of humanoid robots!